Boiler Water Treatment Principles and Practice: Charts and Notes for Field Use

Boiler Water Treatment Principles and Practice: Charts and Notes for Field Use

ISBN: 978-0-8206-0173-1

First Edition:
Chemical Publishing Company, Inc. – 2013

Chemical Publishing Company:
www.chemical-publishing.com

Printed in the United States of America

Table of Contents

About the Author

Colin Frayne, CSci, CChem, CEnv, CWEM, FRSC, FICorr, MCIWEM, MWMSoc (UK), CWT (USA), is an international water treatment consultant and small business owner.

He has more than 45 years of experience in the practice of industrial chemistry and industrial water systems management, and has worked and lectured in over 50 countries. During those years he has also lived on four continents, with his family, while being variously employed in Q.C. and R&D laboratories, in technical sales, sales management, marketing, training, international business development, import/export, and general management.

He graduated in analytical chemistry from North London Polytechnic (now the University of North London), in the United Kingdom, and later obtained various business diplomas from colleges in the U.K. and South Africa, including Wits Business School in Johannesburg.

In November 2004, Colin was presented with the Association of water Technologies, Ray Baum Memorial Award, as "Water Technologist of the Year".

Mr. Frayne holds dual nationality (Britain/USA) and currently lives in Georgia (after spells in New York city and Florida), where he continues to undertakes technology, marketing, and business consulting services work throughout the U.S. and internationally, via his small private company – "Aquassurance, Inc."

He can be reached on +1-561-267-4381,
or by email: aquassurance@msn.com

Summary of Charts and Notes for Field Use

A boiler is a device for heating water or other liquids and is most commonly constructed as a closed pressure-vessel, containing a furnace area, purpose designed heat transfer surface areas, and other functional components. Boilers are commonly used for space heating and domestic hot-water heating purposes, for residential, commercial and institutional markets. However, for all power generation and most industrial process purposes, boilers are employed to generate steam which is achieved by transforming liquid feedwater (FW) into gaseous steam, and so the term "boiler" is commonly interchangeable with the term "steam generator". The objective of generating steam from boiling water is to provide a convenient and (relatively) efficient means of transporting heat energy for subsequent consumption in a particular process, or for conversion to useful work.

Heat Capacity of Water

- Btu is a British Thermal Unit. It is the amount of energy required to raise the temperature of 1 lb water by 1°F
- Heating water from freezing point to boiling point at sea level requires 212–32 = 180 Btu
- Enthalpy of evaporation (*latent heat of vaporization*) = 970 Btu/lb. So, the heat content of steam from and at 212°F = 180 + 970 = 1150 Btu
- Atmospheric pressure = 14.7 psi_{atm} (Approx 1 Bar)

- Absolute pressure (psia) = psi_{atm} + psig
- Also: 1 Btu/lb = 4.177 kJ/kg
- 970 Btu/lb = 2257 kJ/kg
- Latent Heat of Vaporization of water = 1150 Btu/lb (2260 kJ/kg, 593 cal/g) @ 100°C
- Latent Heat of Fusion of water = 144 Btu/lb (334 kJ/kg, 79.7 cal/g) at 0°C

Saturated steam temperatures at various boiler pressures

Saturated steam pressure psig	Absolute pressure psia	Bar absolute	Approximate Temperature °F	Approximate Temperature °C
0	14.7	1.01	212	100
50	64.7	4.46	298	148
100	114.7	7.91	338	170
150	164.7	11.36	366	186
200	214.7	14.80	388	198
300	314.7	21.70	422	217
600	614.7	42.38	489	254
900	914.7	63.07	534	279
1200	1214.7	83.75	569	298
1500	1514.7	104.4	598	314

NOTE: 1 std. atm = 14.696 pounds per square inch absolute (psia or lbf/in^2). 1 bar = 14.5 psia, or 1 kg/cm^2, or 100,000 Pascals (100 kPa)

Boiler Energy and Power Units

Boilers can be classified, rated, sized and compared by means of several energy and power terms. All boilers have a nameplate indicating net and gross output which may be variously indicated in Btu/hr, horsepower (hp), pounds of steam per hour (lb./hr), or square feet of heating surface for any given pressure and temperature listing.

- 1 British Thermal Unit (Btu) = 778.16 ft-lb.
- 1 kiloWatt per hour (kW/h) = 3413 Btu
- 1 horsepower per hour (hph) = 2545 Btu
- 1 calorie = 4.18 Joules (J)
- 1 horsepower (hp) = 550 ft-lb./s
- 1 kW = 1.34 hp
- 1 kilo calorie per min (kcal/min) = 4186 J/min
- 1 therm = 100,000 Btu
- 1 kiloJoule (kJ) = 0.948 Btu
- 1 boiler hp = 33,472 Btu/hr

Typical gross heating values of common fuels (based on approximately 80% fuel to steam efficiency)

- Petroleum coke = 11,500–15,000 Btu/lb. (6,400–8,350 kcal/kg, 26,750–34,900 kJ/kg).
- Anthracite coal = ±14,000 Btu/lb. (7,800 kcal/kg).
- Bituminous coal = 11,500–14,000 Btu/lb.
- Sub-bituminous coal (black lignite) = 8,500–11,500 Btu/lb.
- Lignite = 7,000–9,000 Btu/lb.
- Peat = 5,000–7,000 Btu/lb.
- Bagasse = 3,500–4,500 Btu/lb.
- Wood = 2,500–3,000 Btu/lb. when wet, rising to 8000–9000 Btu/lb. when dry.
- No 1 fuel oil (distillate or kerosene) = ±19,850 Btu/lb. (137,000–138,000 Btu/US gal).
- No 2 fuel oil = ±19,500 Btu/lb. (141,000–142,000 Btu/US gal., or 0.295 US gal. per boiler hp/hr).
- No 5 fuel oil (light residual) = 18,500–19,000 Btu/lb. (148,000–149,000 Btu/US gal., or 0.280 US gal. per boiler hp/hr).
- No 6 fuel oil (residual oil) = 18,000–19,000 Btu/lb. (10,000–10,500 kcal/kg), or about 150,000–152,000 Btu/US gal. (10×10^6 kcal/kg), or approx. 0.275 US gal. per boiler hp/hr.
- Natural gas = 1024–1032 Btu/cu ft, or 40.9 cu ft per boiler hp/hr.
- 500 Btu/cu ft gas = 84 cu ft. gas consumed per boiler hp/hr.

Typical energy consumption and output ratings for a fire tube boiler

Energy/power parameter	Rating
Boiler horse power:	500
Pounds of steam *from and at* 212°F:	17,250
Output: 1000 Btu/hr	16,735
EDR steam, sq. ft: (gross rating)	69,500
EDR steam, sq. ft: (hp and ACC net)	57,250
Approximate oil consumption rate: gph @ 80% efficiency	150
Approximate gas consumption rate: therms per hour @ 80% efficiency	210

Steam is widely employed in all manner of buildings and factories, including:

- Commercial premises and institutions
- Food and beverage industries
- Wood pulp and paper manufacturing
- Chemical and pharmaceutical
- Oil, gas and petrochemicals
- Textiles, dye-houses and laundries
- Metal refining and fabrication
- General manufacturing industry
- Utilities

Steam may be directly indirectly used for innumerable purposes, including:

- Space heating equipment
- Process air heaters
- Evaporators
- Turbines
- Retort cookers
- Jacketed kettles

- Heat exchangers
- Submerged coils
- Dryers
- Strippers
- Hydrostatic sterilizers
- Trace heating lines
- Absorption chillers

The supply and distribution steam systems needs to be meet several basic requirements, including:

1. Vary the steam delivery volumes and pressures to match steam demand.
2. Provide and maintain appropriate high heat-content steam.
3. Ensure the preservation of steam quality (minimum moisture content).
4. Ensure the preservation of appropriate steam purity (minimum of other contaminants).

Steam tables suitable for pressure deaerators

Temp	Temp	Absolute	Bar	Steam	Enthalpy (Btu/lb.)		
°F	°C	pressure psia	bara	pressure psig	Water H_f	Evap H_{fg}	Steam H_g
212	100.0	14.7	1.01	0	180.2	970.3	1150.5
216	102.2	15.9	1.10	1.2	184.2	967.8	1152.0
220	104.4	17.2	1.18	2.5	188.2	965.2	1153.4
224	106.7	18.6	1.28	3.9	192.3	962.6	1154.8
228	108.9	20.0	1.38	5.3	196.3	960.0	1156.3
230	110.0	20.8	1.43	6.1	198.4	958.7	1157.1
232	111.1	21.6	1.49	6.9	200.4	957.4	1157.8
236	113.3	23.2	1.60	8.5	204.4	954.8	1159.2
240	115.6	25.0	1.72	10.3	208.5	952.1	1160.6
244	117.8	26.8	1.85	12.1	212.5	949.5	1162.0
248	120.0	28.8	1.96	14.1	216.6	946.8	1163.4
252	122.2	30.9	2.13	16.2	220.6	944.1	1164.7

NOTE: H_f = enthalpy of saturated water, H_{fg} = enthalpy of vaporization, H_g = enthalpy of saturated steam

Calculating Blowdown

The required boiler BD rate can be expressed in several ways, with some examples given below.

1. Percentage of BW evaporated required to be blowndown: % BW BD = FW TDS × 100/(BW TDS–FW TDS)
2. The percentage of FW supplied that will be lost as BD: % FW BD = FW TDS × 100/BW TDS
3. That part of steam generated lost as BD, in order to maintain BW TDS. BD = Steam generation lb./hr (or kg/hr)/(COC–1)

Where:

TDS is expressed in ppm or mg/l.
COC = cycles of concentration of BW TDS to FW TDS.

Example:

- Boiler steam generation rate is 10,000 lb./hr
- FW TDS is 200 ppm
- BW TDS is maintained at 3000 ppm
- COC = 3000/200 = 15×

Using expression 1. % BW blowdown = 200 × 100/2800 = 7.14%
Using expression 2. % FW blowdown = 200 × 100/3000 = 6.67%
Using expression 3. BD = 10,000/(15–1) = 714 lb./hr

Coefficients of thermal conductivity for some heat-exchanger metals and boiler deposits

Material	Btu·ft./h ft.^{2} °F	W/m^2 °C
Alloy steel	21	36
Carbon steel	32	55
Copper	240	420

Material	Btu·ft./h ft.^{2} °F	W/m^2 °C
Analcite	0.76	1.31
Calcium carbonate	0.56	0.97
Calcium phosphate	2.20	3.81
Ferric oxide	0.35	0.61
Magnetite	1.80	3.11
Serpentine	0.63	1.09

Types of water or steam commonly employed in most HW heating and steam generating plants:

- Raw water (city, potable, reclaimed, treated WW)
- Externally conditioned, preheated, or intermediately treated MU water.
- Finally conditioned or treated MU water.
- Chemical treatment batch dilution water.
- Feedwater.
- Boiler water.
- Boiler water blowdown.
- Steam/water recovered from a FSHR system.
- Steam/water recovered from a process reaction.
- Circulated HW.
- Generated steam.
- Superheated and reheated steam.
- Condensed steam
- Save-all systems water.
- Filtered or polished condensate or return HW.
- Attemperator water.

Commonly occurring minerals in natural MU water sources

Calcium bicarbonate	$Ca(HCO_3)_2$
Calcium chloride	$CaCl_2$
Calcium sulfate	$CaSO_4$
Ferrous bicarbonate	$Fe(HCO_3)_2$
Magnesium bicarbonate	$Mg(HCO_3)_2$
Magnesium chloride	$MgCl_2$
Magnesium sulfate	$MgSO_4$
Manganous bicarbonate	$Mn(HCO_3)_2$
Silica	SiO_2
Silicic acid	H_2SiO_3
Sodium silicate	Na_2SiO_3
Sodium bicarbonate	$NaHCO_3$
Sodium chloride	$NaCl$
Sodium sulfate	Na_2SO_4

Specific waterside/steamside problems affecting MPHW and HPHW boiler plants

Type of problem	Notes
• *Lack of adequate of pretreatment*	Higher temperatures/pressures demands better pretreatment, feed and control. – Use softeners and/or dealkalization for MPHW. – Use demin/RO for HTHW.
• *Prevention of water loss and air ingress*	Good control over water losses and DO ingress is needed to prevent loss of inhibitor and reduced waterside protection. – Use catalyzed sulfite with MPHW. – Use DEHA for HTHW.
• *Control of foulants and corrosion debris*	Thorough pre-cleaning needed to prevent initial fouling in large tortuous systems. Prevent stagnation and glycol degradation. – Use both general and iron dispersants. – Test for sludge/iron. – Side-stream filters very useful.

Salt concentration indicators

Sodium Chloride %W/W NaCl	Specific Gravity	°Beaume	Salometer Degrees	Sodium Chloride Lb./gallon
2	1.014	2.0	8	8.462
4	1.028	4.0	16	8.579
6	1.043	6.0	23	8.704
8	1.059	8.0	30	8.837
10	1.074	10.0	38	8.963
12	1.090	11.9	46	9.096
14	1.107	13.7	53	9.238
16	1.124	15.5	61	9.380
18	1.142	17.3	68	9.530
20	1.160	19.1	76	9.680
22	1.179	20.8	84	9.839
24	1.197	22.5	81	9.989
26	1.204	24.4	99	10.047

Summary of waterside/steamside problems affecting LPHW and LP steam heating boiler plants

Type of problem	Notes
1. Problems with heating coils • *Internal coil corrosion* **Note:** corrosion debris is: *green* hydrated copper carbonate $Cu^{[II]}CO_3 \cdot nH_2O$ *red* cuprous oxide Cu_2O	**Acid corrosion** from soft water. **Pinhole corrosion** from O_2 and CO_2. **Erosion corrosion** over 6 ft/s flow.
• *Internal coil deposition*	**Hard water scale** from hard water. *White* calcium carbonate $CaCO_3$ **Phosphate scale** from threshold agents. *blue/green/black* copper phosphate trihydrate $Cu_3(PO_4)_2 \cdot 3H_2O$ **Dirt pockets** due to low level of coil.
• *External coil problems*	**BW sludge** settles on external fins. **Gasket weeping** from inhibitor. **Color stains** from inhibitor dyes.
2. Problems with glycols • *Loss of glycol strength*	**Reduced freeze protection** from water leaks and glycol oxidation.
• *Glycol degradation*	**Acid corrosion** (system pH < 7) from extreme degradation of glycols.
3. Problems of corrosion • Oxygen *corrosion* **Note:** corrosion debris is: *red/brown* hematite sludge *black* coarse magnetite	**Tuberculation** and **pitting** due to DO, air in-leakage, loss of nitrite and lack of specific oxygen scavengers.
• *Galvanic & under-deposit corrosion*	**Mixed metal corrosion** Cu/Fe/Zn **Iron transport** -iron sludges, esp. ferrous bicarbonate $Fe(HCO_3)_2$ **UDC** from fouling and binders.
4. Other system problems • *Uncontrolled* and *excessive water losses*	**Pump**, **valve**, **steam trap losses** and no water meter leads to O_2 pitting, Fe fouling and deposits of carbonate scale and ferrous metasilicate $FeSiO_3$
• *Lack of BD control*	**Water losses, low inhibitor levels**
• *High treatment/small boiler*	**Surging/carryover** from high TDS.
• *Poor design, lead and lag boiler operation*	**Excessive TDS** in lead boiler. **Lack of treatment** in lag boiler.

FW contamination from MU water

Type of problem	Likely results or increased risks
Lack of ion-exchange softener with low hardness MU water source	• Extra chemical treatment need • Waterside fouling • Baked on sludge • Increased BD and higher costs
Lack of dealkalization with high alkalinity MU water **Note:** Anion dealkalizers do not use acid, but also do not reduce MU water TDS and introduce additional chlorides (risk of pitting).	• Excessive CR line corrosion • BW carryover • Low cycles of concentration • Higher operating costs
Hardness breakthrough with ion-exchange (base exchange, BX) softening **Note:** Caused by Fe/Mn fouling, resin breakdown/loss, or inadequate regeneration.	• Increased risk of carbonate scale or phosphate sludge • Loss of alkalinity and hence an increased silica deposition risk
Chloride leakage with BX softening **Note:** Caused by inadequate rinsing.	• BW carryover • Increased BD • Temporary loss of control • Depassivation
Acid breakthrough (split-stream softening, dealk./DI).	• Failure to neutralize results in acid corrosion of feed system
Na leakage (DI). **Note:** Deposits may develop containing high concentration of NaOH, resulting in:	• Caustic gouging • Caustic induced SCC (where stressed)
Silica leakage (DI).	• Silica volatilization • Deposition on boiler surfaces, superheaters & turbine blades
Cl^- and SO_4^{2+} leakage (DI).	• Anodic depolarization, leading to depassivation and corrosion
Organic fouling of DI resins.	• Loss of treated water capacity • Anion leakage.
External treatment carryover, and after-precipitation.	• Fouling and blocked check valves • Erosion/corrosion

FW contamination from returned condensate

Type of problem	Likely results or increased risks
Carrythrough of BW solids with the condensate.	• Highly alkaline condensate • Pre-boiler metal damage • pH control out of balance • Fouling to DA vent, check valves and FW pump
High O_2 loading of returning condensate.	• High deaeration demand that may exceed DA capacity.
Steam/CR system corrosion debris pickup, transport and re-deposition.	• Wide ranging FW line blockages and impingement/erosion • Enhances risk of pre-boiler corrosion
Process contamination.	• Sticky films (LP boilers) • Varnish (HP boilers) • Steam discoloration • FW pH fluctuations • Acid corrosion • Stable foams leading to carryover
Oil and hydrocarbon leaks.	• FW system fouling and deposit binding • Non-wettable boiler surfaces.
Condenser leaks.	• Fouling, deposition • Chloride depassivation and pitting corrosion to FW lines and boiler

Problems associated with the final FW blend

Type of problem	Likely results or increased risks
Inadequate FW deaeration.	• Oxygen pitting corrosion to deaerator, economizer, FW tank, FW line • Pitting in tubes and shell (FT) • Pitting to top drum (WT)
Dealloying, selective leaching and graphitic corrosion forms of galvanic corrosion. **Note:** Corrosion is especially common in soft, waters, with high chlorides and under idle conditions.	Selective corrosion of cuprous alloys, such as FW heaters and bronze pump impellors, producing: • Dezincification • Denickelification • Destannification, • Dealuminumification. • Graphitic corrosion of cast iron FW pumps and valves.
High temperature/stress and stop/start operation effects on 70:30 cupronickel tubes (as found in some FW heaters).	A chain of cause and effect including: • Oxygen corrosion • Dealloying • Exfoliation corrosion.
Impingement by high speed, turbulent water, and particles.	• Erosion-corrosion of pumps impellers, fittings.
Air bubble entrainment in FW.	• FW pump cavitation
Localized pre-boiler scale and corrosion debris deposits.	Combination of: • *New* phosphate, iron, copper, and silica deposition • *Old* re-deposited debris • Transport of Fe, Cu, Ni, Zn, Cr oxides to HP boiler section, leading to deposition, fouling and possible tube failures • Transport of minerals and debris including malachite, ammonium carbamate, basic ferric ammonium carbonate • Precipitation in FW line of phosphates, iron and silicates

Deposition of boiler section waterside surfaces by alkaline earth metal salts, other inorganic salts and organics (Note: Deposition can also take place in the pre-boiler section)

Deposit material	Mineral/formula	Problem or risk
Calcium carbonate	*Aragonite* $CaCO_3$	Low temperature scale.
	Calcite $CaCO_3$	Higher temperature scale.
Calcium phosphate	*Hydroxyapatite* $Ca_{10}(OH)_2(PO_4)_6$	Flocculated boiler sludge, as a result of residual hardness.
	Tri-calcium phosphate $Ca_3(PO_4)_2$	Calcium phosphate can form as a hard, adherent scale in FW line, especially if pH is below 8.2–8.3.
Calcium sulfate	*Anhydrite* $CaSO_4$	Hard, difficult scale.
(Basic) Magnesium chloride	*Magnesium chloride hydrate* $MgCl_2 \cdot 5Mg(OH)_2 \cdot 8H_2O$	Saline concentrates in porous deposits, resulting in acid corrosion pits.
Magnesium hydroxide	*Brucite* $Mg(OH)_4$	Common flocculated boiler sludge.
(Basic) Magnesium phosphate	*Magnesium hydroxy-phosphate* $3Mg_3(PO_4)_2 \cdot Mg(OH)_2$	Forms with hematite in passivated films at lower temperatures.
Sodium hydroxide	*Caustic soda* NaOH	Caustic concentrates in porous deposits, resulting in caustic gouging and SCC.
Various sludges	*Phosphates, etc.*	Baked on sludge.
Treatment chemicals	*Sulfites, hydrazine, amines, chelants*	Incorrect application results in thermal breakdown, sludges and corrosion deposits.
	Tannins, dispersants, hydrocarbons	Carbonization produces deposits and binders.

Silica and silicate crystalline scales and deposits affecting boiler section waterside surfaces

Mineral/Formula	Problem or risk
Acmite $Na_2O{\cdot}Fe_2O_3{\cdot}4SiO_2$	High temperature scale. Found under sludges of hydroxyapatite, serpentine or porous iron oxides. Can result from alum carryover.
Alpha quartz αSiO_2	Hard scale found in drum, tubes and turbine blades.
Analcite $Na_2O{\cdot}Al_2O_3{\cdot}4SiO_2{\cdot}2H_2O$	Similar to acmite. Presence can indicate steam blanketing problems.
Magnesium orthodisilicate $Mg_3Si_2O_7{\cdot}2H_2O$	Sludge forms from magnesium hydroxide and silicate in boiler.
Natrolite $Na_2O{\cdot}Al_2O_3{\cdot}3SiO_2{\cdot}2H_2O$	Moderate hardness scale.
Nepheline various structures eg: $[Na_4K_2Al_6(SiO_4)_2]$	Moderate hardness scale.
Noselite $5Na_2O{\cdot}3Al_2O_3{\cdot}6SiO_2{\cdot}2H_2O$	Hard, adherent scale found on tubes and high heat-flux density areas.
Serpentine $3MgO{\cdot}2SiO_2{\cdot}H_2O$	Common sludge, forms from Mg salts and silicate at lower pressures.
Sodalite $3Na_2O{\cdot}3Al_2O_3{\cdot}6SiO_2{\cdot}2NaCl$	Hard adherent scale on tube surfaces.
Wollastonite $CaSiO_3$	Moderate hardness scale.
Xonotlite $5CaO{\cdot}5SiO_2{\cdot}H_2O$	Hard, adherent scale found on tubes and high heat-flux density areas.

Iron oxide and other boiler section corrosion debris deposits

Mineral/Formula	Problem or risk
Ammonium carbamate $NH_2Fe_2(OH)_4(CO_3)_2 \cdot H_2O$ *Hydrated basic ferric ammonium carbonate* $(NH_4)2Fe_2(OH)_4(CO_3)_2 \cdot H_2O$ *Malachite* $CuCO_3 \cdot Cu(OH)_2$	The transport of pre-boiler corrosion debris to the boiler section will include these minerals. Also the oxides of Fe, Cu, Ni, Zn and Cr as noted below.
Bunsenite NiO	Nickel oxides. A result of corrosion of brasses and cupronickels.
Copper Cu	Deposited by direct exchange with iron, or by reduction of copper oxide, during the corrosion of steel.
Cuprite CuO	Black non-magnetic mineral. Originates from condensers and FW heaters.
Eskolaite Cr_2O_3	Chromic oxide. A result of corrosion of stainless steels.
Ferric phosphate $FePO_4$	Found in low temperature passivation film with hematite.
Ferrous bicarbonate $[Fe(HCO_3)_2]$	Transport into FW line and boiler. Common in smaller boilers with a high percentage of cold water MU water.
Haematite Fe_2O_3	Formed at lower temperatures and higher oxygen concentrations.
Magnetite Fe_3O_4	Formed at high temperatures, under fully reducing conditions.
Siderite $FeCO_3$	Formed in LP boilers with high alkalinity BW.
Wustite FeO	Ferrous oxide formed in LP boilers with poor DO removal. Develops as the core of tuberculation deposits.
Zincite ZnO	Originates from *brasses* used for condensers and FW heaters.

Boiler section corrosion problems involving oxygen, concentration cells and low pH

Type of problem	Notes
Coarse magnetic corrosion Fe_3O_4 **Note:** Passivation is a form of corrosion, albeit the resulting magnetite is desirable.	Thick, porous, bulk water or boiler surfaces deposit. Formed under both high and low pH and other adverse BW conditions. Of no practical passivation benefit.
***Oxygen corrosion**, occurring as:* • *General etch corrosion/uniform rate corrosion (less common form)* • *Localized corrosion (takes several forms and common where waterside conditions are less-than-ideal).*	Rare in correctly treated boilers but can affect drum waterline and tubes. More common in idle and low-load boilers, where water chemistry is unbalanced, under high MU conditions and after poor chemical cleaning. Also in peak-load boilers, especially where deposition can occur. Localized corrosion can be very serious, causing metal failure.
***Localized, concentration-cell corrosion (differential aeration corrosion)**, occurring as:* • *Tuberculation corrosion* • *Crevice corrosion* • *Under-deposit corrosion* • *Pitting corrosion*	All forms of localized, concentration-cell corrosion are "indirect attack" type corrosion mechanisms. They result in severe metal wastage and can also induce other corrosion mechanisms, e.g. • Stress corrosion • Corrosion fatigue
*Other forms of **concentration-cell corrosion** include:* • *Caustic gouging* • *Saline corrosion*	Combination of free caustic and concentrating effect causes severe metal wastage. High chlorides and sulfates, result in corrosion from depolarization and depassivation effects
***Low pH corrosion** includes:* • *Low pH general corrosion* • *Low pH localized corrosion* • *Acid cleaning corrosion*	– Results from acid breakthrough into boiler water with only limited buffering capacity. – Requires an acid source and a concentration effect. – Results from both poor cleaning and poor neutralization processes.

Stress and high temperature related corrosion

Type of problem	Notes
Embrittlement corrosion *occurring as:* • *caustic cracking (caustic embrittlement)* • *hydrogen embrittlement (hydrogen damage)*	Also known as **stress corrosion cracking** (SCC) – Intergranular corrosion affects both carbon steels and austenitic steels and accelerated by high stress, higher temperatures and impurities in grain boundaries. – Only in boilers over 1000 psig, leading to violent tube rupture.
Fatigue corrosion *occurring as:* • *Thermal fatigue cracking (thermal effect corrosion)* • *Corrosion fatigue*	– Cycles of thermally induced stress leads to metal failure. – Results from a combination of thermal cycling stress and SCC or other corrosion process.
High temperature corrosion *occurring as:* • Long-term overheating • Short-term overheating • High temperature corrosion • Thermal oxidation (metal burning/metal scaling) • Spheroidization • Decarburization • Graphitization	Effects of overheating lead to premature metal failure through: – Bulges – Magnetite blisters – Fractures – Creep/ductile metal deformation – Graphitization decomposition – Fish mouth ruptures – Spalling of flakes/splinters – Exfoliation of surface chips
Chelant corrosion (***chelant attack***)	Complexing of soluble Fe, Cu and Cu alloy ions. (Excess chelant and oxygen leads to change in redox potential and corrosion develops).
Copper corrosion occurring as: • *Transported corrosion debris* • *Cupric ammonia complex ion corrosion* • *Anodic area pitting corrosion* • *Liquid metal embrittlement*	– Resulting from ammoniacal corrosion of steam/CR lines. – Resulting plated copper acts as cathode to surrounding anodic boiler steel, inducing corrosion. – Results from poor acid cleaning where copper traces are present. – Results from copper deposits and high temperature conditions.

Steam purity, quality and other operational problems

Type of problem	Results and notes
Poor steam purity	Contaminated steam leads to corrosion, erosion, sticky valves and boiler operation problems.
Poor steam quality	Inadequate steam/water separation results in "wet steam" with reduced heat content.
Poor steam sampling (common in smaller boiler plant)	Lack of sampling facilities and/or poor protocols results in limited awareness of steam related problems and risky conditions.
Poor matching of steam volumes, pressures, temperatures and heat-content with demand.	A failure to match steam output with demand impacts on both process and waterside conditions.
Corrosive gases including: • *oxygen* • *carbon dioxide* • *ammonia* • H_2S/SO_2	Results in: – Oxygen corrosion. – Carbonic acid corrosion. Greatly enhanced when O_2 present. – Ammoniacal corrosion. – Sulfurous acid corrosion.
Vaporous silica present as *silicic acid* and *silicate ion*	• Crystalline silica deposits • Amorphous silica deposits
Scale and corrosion debris transport results in:	• FW line/feed pump blockages • Impingement and erosion • Poor deaerator venting • Sticking check valves • Fe,Cu,Ni,Zn,Cr oxides • Crystalline magnetite
Oil, fat, grease and hydrocarbon contamination results in:	• Film boiling • Coke deposition • Contaminant binding • Sludging and gunk balls
Industrial process contamination results in:	• Cross-contamination • Malodors and discoloration • Sticky films and varnishes • Stable foams • Chemical incompatability • Impingement corrosion • Fluctuating pH/acid corrosion

Specification for grades of high-quality water suitable for higher pressure WT boilers.

Boiler rating	**600 to 900 psig (41 to 62 bar)**	**900 to 1500 psig (62 to 103 bar)**	**over 1500 psig (over 103 bar)**
MU water quality	**Basic pure water**	**Very pure water**	**Ultrapure water**
Resistivity MΩ/cm at 25°C	Above 0.05 for 90% of time. Not less than 0.02	Above 1 for 90% of time. Not less than 0.1	Above 15 for 90% of time. Not less than 10
Non-reactive silica µg/l SiO_2	50 to100	10 to 20	5 to 10
Sodium µg/l Na	5000 to 10000	100 to 250	1 to 10
Total organic carbon µg/l TOC	300 to 500	50 to 100	<50

NOTE: 0.05 MΩ/cm = 20 µS/cm, 1 MΩ/cm = 1 µS/cm, 15 MΩ/cm = 0.066 µS/cm. Practical resistivity limit for ultrapure water is 18.3 MΩ/cm at 25°C (0.055 µS/cm at 25°C). Also: 1000 µg/l = 1 mg/l = 1 ppm.

Some practical considerations for a RW ion-exchange softener are:

- Resin type is strong acid cation in sodium form, fully hydrated; typically 53 lb/cu ft at 45% moisture content (0.85 kg/l).
- For continuous softening service it is advisable to install two or more softeners in parallel, on a duty/standby basis.
- The minimum water pressure required is typically 20 psig.
- Total resin volume design requirement is sufficient resin for 8 hours minimum continuous operation; 12 to 16 hours per tank is fine.
- Typical resin exchange capacity is 20,000 grains as $CaCO_3$/cu ft resin (45.9 grams hardness/l resin) at 6 lb. NaCl/cu ft of resin (0.1 kg/l) to 30,000 grains at 15 lb. NaCl/cu ft. Resin capacity decreases and calcium leakage increases with high sodium to calcium ratios.

- Typical salt brine regenerant concentration is 10 to 25% NaCl; maximum is 36%. Caution is needed when using some types of rock salt, as it may contain considerable calcium and magnesium salts. Good quality rock salt is 98 to 98.5% NaCl. Evaporated, re-crystallized salt is 99.5 to 99.8% NaCl.
- Resin bed depth is 24 in minimum, 72 in maximum (0.61 to 1.83 m).
- Bed expansion (freeboard) is 50% minimum (thus the resin tank needs to be at least double the volume of the resin requirement). Resin bed expansion is a function of backwash rates and temperature.
- Service flow rate is 15 to 30 bed volumes (BV)/hour, or 2 to 4 gpm/cu ft resin (0.27 to 0.40 lpm/l resin). Linear flow at 4 to 10 gpm/cu ft.
- Regeneration: Backwash rate is 5 to 6 gpm/cu ft for 10 min (0.69 to 0.80 lpm/l resin) at 70°F (21°C) and 10 gpm/cu ft (1.38 lpm/l resin) at 190°F (88°C).
- Regeneration: Typically, brine injection rate is 0.5 to 1 gpm NaCl soln/cu ft resin (0.13 lpm/l resin). Typically, this takes 25 to 30 minutes. Check NaCl strength. It may take 4 to 16 hours to achieve suitable strength.
- Regeneration: Slow Rinse is 1 gpm water/cu ft resin for 15 minutes (0.13 lpm/l resin).
- Regeneration: Fast Rinse is 1.5 gpm water/cu ft resin for 5 minutes (0.2 lpm/l resin).

Types of Internal Treatment Program

Internal treatments for BW are chemical formulations having potentially beneficial single-function, dual-function or multifunctional effects. Globally, there are many thousands of BW formulations available, and for the most part, they can each be loosely classified into one or other type of commonly recognized chemical program, as described below.

- Anodic inhibitor programs: These programs are based on ingredients such as nitrite, silicate and molybdate chemistries and are usually formulated as light-duty multifunctional programs in HW heating and LP steam boiler systems.
- Tannin programs: These programs are based on blends of certain natural or synthetic tannins and provide oxygen scavenging, passivating and sludge conditioning functions. They can be formulated into complete multifunctional programs or simple blends that are supported by other functional chemicals or programs. Tannin programs can be employed in a wide range of commercial and industrial boiler types and pressures.
- Coagulation and precipitation programs: These programs are extremely widely used and employ various types of phosphates as a precipitant to provide control over the unwelcome deposition of hardness scales. Carbonate and polysilicates were once commonly used, but less so today.
- Chelant programs: These programs are commonly prescribed for both FT and WT boilers and are employed either as replacements for, or used in combination with, phosphate precipitation programs.
- All-polymer programs: There are many types of all-polymer programs and today, with the wide availability of many specialty polymers, they are proposed for application in almost every type of boiler from the smallest to the largest. Typically they are multifunctional, replacing precipitation programs and sludge conditioners, and providing control over specific problems such as iron transport that can affect many boiler plant facilities. In practice, many all-polymer programs are more properly called all-organic programs as they may contain organics, such as phosphonates, that are not necessarily polymers.

- Chelant, phosphate or polymer based, dual or multifunctional programs: Every water treatment service company offers several types of dual- or multifunctional program based on either chelant or phosphate, together with polymers and related organics. The various combinations of chemistry, ratios of primary ingredients employed and activity levels offered have given rise to an enormous number of permutations. Some early program examples included carbonate/polymer, phosphate/carbonate, chelant/phosphate and chelant/phosphonate. The development of more modern polymers led to chelant/polymer and phosphate/polymer, and multifunctionals such as chelant/polymer/phosphate and alkali/polymer/phosphate.
- Coordinated phosphate programs: These are programs formulated to avoid the formation of hydroxyl alkalinity (free caustic). They tend to be employed in higher-pressure WT boilers and require very careful control. Variants include congruent control and equilibrium phosphate treatment.
- All-volatile programs (AVPs): These programs are employed in higher-pressure boilers (generally power boilers) and utilize only volatile chemicals, such as ammonia, amines (such as diethylhydroxylamine DEHA), and other vapor phase inhibitors (VPIs).

In addition to full- or part-treatment formulations (boiler compounds), the art and science of BW treatment usually requires the application of certain primary functional chemicals and secondary functional chemicals (adjuncts) to provide a comprehensive program, as outlined below:

- Program primary support chemicals encompass oxygen scavengers and condensate treatments.

  Oxygen scavengers include inorganics such as sodium sulfite and organics such as hydrazine.

  Condensate treatments include neutralizing amines such as morpholine and filming amines such as octadecylamine (ODA).
- Adjuncts include antifoams/defoamers, some problem-specific polymers and alkalinity boosters.

Antifoams control foam after it has developed and defoamers are preventative products. Natural products such as castor oil and linseed oil were long ago replaced by synthetics including polysilicones and polyethoxylates. Problem-specific polymers include terpolymers, used for iron dispersion, iron transport and silica control. Alkalinity boosters are typically based on potassium hydroxide or sodium hydroxide.

Carbonate Cycle Requirement Calculations

This table provides a useful starting point for establishing a carbonate BW reserve, but fails to properly address the problem of rapid breakdown of carbonate at increasingly higher pressures, such that the desired carbonate residual may never be reached.

FW alkalinity requirements using 100% soda ash can be calculated using the formula below.

$$\left[0.75 \times \text{FWTH} + \left(\frac{\text{RBWMA}}{\text{COC}} - \text{FWMA}\right)\right]$$

$= \text{lb. of product required, per } 1 \times 10^6 \text{ lb. of boiler FW}$

Where:

FWTH = FW total hardness, in mg/l (ppm) $CaCO_3$
FWMA = FW methyl orange/total alkalinity, in mg/l $CaCO_3$
RBWMA = Required BW methyl orange alkalinity, in mg/l $CaCO_3$
COC = Cycles of FW concentration present/desired in the BW

In summary, the carbonate-cycle program provides preferred precipitation and coagulation reactions to prevent hard scale from forming. Key functions are:

- Calcium hardness precipitates as calcium carbonate sludge
- Magnesium salts precipitate as the hydroxide or silicate

- Calcium sulfate is deposition is inhibited by the addition of sufficient soda ash

Phosphate-Cycle Requirement Calculations

For precipitating phosphate programs it is necessary to ensure adequate boiler water (BW) alkalinity in addition to the phosphate reserve. Sodium hydroxide should be used.

The alkalinity feed-rate requirement is based on FW consumption and calculated as follows:

$$\left[0.67 \times FWCaH + \left(FWMgH - FWMA + \frac{RBWPA}{COC}\right)\right] \times \frac{100}{\%\ NaOH\ in\ product}$$

= lb. of product required, per 1×10^6 lb. of boiler FW

Where:

FWCaH = FW calcium hardness, in mg/l (ppm) $CaCO_3$
FWMgH = FW magnesium hardness, in mg/l $CaCO_3$
FWMA = FW methyl orange/total alkalinity, in mg/l $CaCO_3$
RBWPA = Required BW phenolphthalein alkalinity, in mg/l $CaCO_3$
COC = Cycles of FW concentration required in the BW

A Guide to Tannin Residuals in BW

Boiler pressure		Tannin Reserve	Tannin Index/
psig	bar	ppm	Value
0 to 15	0 to 1	150 to 200	15 to 20
15 to 217	1 to 15	100 to 150	10 to 15
217 to 435	15 to 30	50 to 100	5 to 10
435 to 650	30 to 45	20 to 50	2 to 5

Carbonate-Cycle Program. BW Carbonate Reserve Requirements by Pressure and Sulfate Concentration

Pressure	**100 psig**	**200 psig**	**300 psig**	**400 psig**
Sulfate $(SO_4)^{2-}$ Concentration	Carbonate $(CO_3)^{2-}$ Reserve	Carbonate $(CO_3)^{2-}$ Reserve	Carbonate $(CO_3)^{2-}$ Reserve	Carbonate $(CO_3)^{2-}$ Reserve
100 ppm	5	15	30	45
200 ppm	10	30	55	90
300 ppm	15	45	80	135
400 ppm	20	60	110	185
500 ppm	25	75	140	235
1000 ppm	45	145	280	
1500 ppm	65	210		
2000 ppm	90	285		
2500 ppm	115	350		
3000 ppm	140			
3500 ppm	160			
4000 ppm	180			

Carbonate-Cycle Coagulation and Precipitation Program. Recommended BW Control Limits for Non-Highly-Rated FT Boilers Employing Hard or Partially Softened FW

FW hardness, ppm $CaCO_3$	**40 to 50**		**5 to 10**	
Pressure, psig/bar max. (approximate)	**150/10**	**250/17**	**150/10**	**250/17**
BW chemistry control limits				
TDS, ppm max.	2000	1750	2500	2000
Silica, ppm SiO_2 max.	100	100	100	100
Suspended solids, ppm max.	300	250	200	150
P Alk., ppm $CaCO_3$	30 to 150 in all cases			

NOTE: This chart assumes oxygen scavenger, polymer and other necessary water chemistry controls are in place. If the FW is fully softened, there is no benefit in using this program. Employ a phosphate-cycle or alternative program.

Phosphate-Cycle Coagulation and Precipitation Program. Recommended BW Control Limits for Non-Highly-Rated FT Boilers Employing Hard, Partially Softened, or Fully Softened FW

FW hardness, ppm $CaCO_3$	**40 to 50**		**5 to 10**		**1 to 2**
Pressure, psig/bar max. (approximate)	**150/10**	**250/17**	**150/10**	**250/17**	**350/25**
BW chemistry control limits					
BW TDS, ppm max.	2000	1750	2500	2000	3500
Silica, ppm SiO_2 max.	140	140	140	140	175
S/Solids, ppm max.	300	250	200	150	50
T Alk., ppm $CaCO_3$ max.	700	600	900	700	1200
OH Alk., ppm $CaCO_3$	In all cases this should be 350 min.				
Phosphate, ppm PO_4	In all cases this should be 30 to 60				

NOTE: This chart assumes oxygen scavenger, polymer and other necessary water chemistry controls are in place.

Phosphate-Cycle Coagulation and Precipitation Program. Recommended BW Control Limits for Non-Highly-Rated WT Boilers Employing Hard, Partially Softened, or Fully Softened FW

Pressure, psig/bar max. (approx.)	**350/25**	**650/45**	**950/65**
FW hardness, ppm, $CaCO_3$	**40 to 50**	**10 to 20**	**5 to 10**
BW chemistry control limits			
TDS, ppm max.	2000	2000	1500
Silica, ppm SiO_2 max.	0.4*x* OH alkalinity	0.4*x* OH alkalinity	0.4*x* OH alkalinity
Suspended solids, ppm max.	300	200	100
T Alk., ppm $CaCO_3$ max.	700	600	500
OH Alk., ppm $CaCO_3$ min.	350	300	250
Phosphate, ppm PO_4	30 to 60	30 to 50	20 to 40

Pressure, psig/bar max. (approx.)	**350/25**	**650/45**	**950/65**
FW hardness, ppm, $CaCO_3$	**1 to 2**	**<1.0**	**<0.5**
BW chemistry control limits			
TDS, ppm max.	2500	1500	1000
Silica, ppm SiO_2 max.	0.4*x* OH alkalinity	0.4*x* OH alkalinity	0.4*x* OH alkalinity
Suspended solids, ppm max.	50	20	5
T Alk., ppm $CaCO_3$ max.	600	500	300
OH Alk., ppm $CaCO_3$ min.	300	150	60
Phosphate, ppm PO_4	30 to 60	20 to 40	15 to 30

NOTE: This chart assumes oxygen scavenger, polymer and other necessary water chemistry controls are in place. If determining TDS via a conductivity meter, use an un-neutralized sample.

Chelant demand (ppm product) per 1ppm substrate

Substrate	**38% EDTA**	**Hampene Na_2**	**40% NTA**
Chelant molecular weight	*380*	*372*	*257*
Hardness, ppm $CaCO_3$	10.0	3.7	6.5
Calcium, ppm Ca	25.0	9.3	17.3
Magnesium, ppm Mg	41.0	15.5	28.3
Iron, ppm Fe	17.9	6.6	12.3
Copper, as ppm Cu	15.7	5.8	10.8
Aluminum, as ppm Al	37.1	13.7	25.5

NOTE: ***38% EDTA*** *= 41.6% EDTA $Na_4 \cdot 2H_2O$, Hampene or Versene 100 (100 mg $CaCO_3$/g). Typically supplied as a 40% w/w product.* ***Hampene Na_2*** *= 99% EDTA $Na_2 \cdot 2H_2O$, (267 mg $CaCO_3$/g).* ***40% NTA*** *= 42.8% NTA $Na_3 \cdot H_2O$, Hampene 150 (156 mg $CaCO_3$/g). Typically supplied as a 43% w/w product.*

EDTA Chelant or All-Polymer/All-Organic Program. Recommended BW Control Limits for Fired WT Boilers Employing Demineralized or Similar Quality FW

Pressure, psig/bar max. (approx.)	**350/25**	**650/45**	**950/65**
FW chemistry control limits			
Hardness, ppm $CaCO_3$ (max.)	2	1	0.2
Total Fe/Cu/Ni, ppm (max.)	0.05	0.05	0.03
BW chemistry control limits			
Conductivity, µs/cm/25°C max.	3000	2000	1000
Silica, ppm SiO_2 max.	0.4*x* OH alkalinity	0.4*x* OH alkalinity	0.4*x* OH alkalinity
OH Alk., ppm $CaCO_3$	50 to 250	50 to 100	20 to 50
Polymer, ppm "actives"	15 to 25	10 to 15	5 to 10
OR Chelant reserve, ppm $CaCO_3$	4 to 8	3 to 6	2 to 3

Pressure, psig/bar max. (approx.)	**1250/85**	**1550/105**	**1850/130**
FW chemistry control limits			
Hardness, ppm $CaCO_3$ (max.)	<0.05	ND	ND
Total Fe/Cu/Ni, ppm (max.)	0.02	0.02	0.02
BW chemistry control limits			
Conductivity, µs/cm/25°C max.	250	60 to 80	30 to 40
Silica, ppm SiO_2 max.	8	2.0	1.0
OH Alk., ppm $CaCO_3$	5 to 20	ND	ND
Polymer, ppm "actives"	2 to 5	2 to 3	1 to 2
OR Chelant reserve, ppm $CaCO_3$	1 to 2	unsuitable	unsuitable

NOTE: This chart assumes oxygen scavenger, FW pH and other necessary water chemistry controls are in place.

FW total metals sampled from economizer outlet.

Conductivity test is on un-neutralized samples.

ND = not detected.

Oxygen Solubility at Atmospheric Pressure

Temperature		Oxygen solubility	
°C	°F	mg/l (ppm)	ml/l
25	77	8.5	12.2
30	86	7.3	10.5
35	95	6.7	9.6
40	104	6.2	8.9
45	113	5.9	8.5
50	122	5.4	7.7
55	131	5.0	7.2
60	140	4.6	6.6
65	149	4.1	5.9
70	158	3.8	5.4
75	167	3.1	4.4
80	176	2.7	3.9
85	185	2.2	3.2
90	194	1.7	2.4
95	203	1.0	1.4
100	212	0.0	0.0

NOTE: 0.005–0.01 cc/l O_2 = 0.0072 to 0.0143 part per million (ppm), 0.0072 to 0.0143 milligram per liter [mg/l], or 7.2 to 14.3 parts per billion (ppb).
To convert DO in cc/l to mg/l, multiply by 1.433, for 0°C and 760 mm pressure. Solubility corrections are needed for different temperatures, pressures and salt concentration, but can be ignored for the low levels of DO in deaerated water.

Properties of Oxygen Scavengers

Oxygen Scavenger *Note: Combining ratio is for 100% scavenger*	**Combining Ratio (Practical)**	**Max. Pressure psig**	**Volatility/ Distrib. Ratio**	**Passivation ability**
Sodium sulfite as solid or 32% soln. Cat. = Co or Eryth.	10:1 Rapid scavenger	950 max. Risk of SO_2/H_2S	Not volatile DR = Nil	Limited, only over 300 psig
Sodium bisulfite Typical is 40% soln. Cat. = Co or Eryth	7:1 Rapid scavenger	950 max. Risk of SO_2/H_2S	Not volatile DR = Nil	Limited, only over 300 psig
Na metabisulfate 100% powder Cat. Co or Eryth.	5:1 Rapid scavenger	950 max. Risk of SO_2/H_2S	Not volatile DR = Nil	Limited, only over 300 psig
Hydrazine as 15 or 35% soln. Cat. = HQ	3:1 Rapid scavenger	2500+ Produces NH_3	Poor volatility DR = 0.1	Excellent for all system
DEHA as 17.5 to 30% soln. Cat. = HQ or Cu	3:1 Rapid scavenger	2500+ Some NH_3	Good volatility DR = 1.3	Excellent for all system
Erythorbate as 10 to 20% soln. Cat. = Cu/Ni/Fe	10:1 Good scavenger	1500 Nothing harmful	Not volatile DR = Nil	OK, but only in the boiler
Hydroquinone as 15 to 25% soln. Cat. = pyrogallol	7:1 Scavenging enhancer	1500 Produces CO_2	Only at 1500 psig DR = 0.15	Acts as enhancer only
MEKO Poor solubility Cat. = Erythorbate	6:1 Weak scavenger	1250 Risk of charring	Highly volatile DR = 2.2	No true ability to passivate
Carbohydrazide as 6.5% solution Cat. = Cu or Co	1.5 Slow scavenger	2500+ Produces CO_2/NH_3	Some, but only over 130 psig	Excellent but only utilities
1-Aminopyrollidine as 30% solution Cat. = HQ	2:1 OK as scavenger	1250 Risk of Charring	Some volatility DR = 0.7	OK, but not special
Hydrolysable tannins as 25 or 50% soln. Cat. = None or Eryth.	10:1 Good for cold FW	650 max. Product then fails	Not volatile DR = Nil	Excellent but only in boiler

NOTE: 1 std. atm = 14.696 pounds per sq. inch absolute (psia or lbf/in^2).
1 bar = 14.5 psia, or 1 kg/cm^2, or 100,000 Pascal, or 100 kPa, or 0.1 MPa

Carbon Dioxide Evolution from FW Alkalinity

Boiler pressure	**Bicarbonate alkalinity starting point:** ***1 ppm carbonate (as $CaCO_3$) produces the following ppm of CO_2***	**Carbonate alkalinity starting point:** ***1 ppm carbonate (as $CaCO_3$) produces the following ppm of CO_2***
50 psig	0.62	0.18
100 psig	0.70	0.26
150 psig	0.79	0.35
200 psig	0.84	0.40
250 psig	0.86	0.42
500 psig	0.88	0.44

Amine Requirement to Reach a Stable Condensate pH

ppm CO_2 in steam	**18**	**18**	**32**	**32**	**58**	**58**
Desired condensate pH	*7.5*	*8.0*	*7.5*	*8.0*	*7.5*	*8.0*
AMP: ppm 40% strength reqd.	68	73	155	168	298	312
CHA: ppm 40% strength reqd.	95	103	168	173	310	328
DEAE: ppm 40% strength reqd.	88	103	180	190	325	338
Morph: ppm 40% strength reqd.	93	125	193	313	350	380

Amine Basicity Dissociation Constants

Amine	**MW**	K_b	pK_a/pK_b
Morpholine	87	3.1×10^{-6}	8.4/5.6
Ammonia	17	18×10^{-6}	9.3/4.7
DMIPA	103	40×10^{-6}	9.5/4.5
DEAE	117	45×10^{-6}	9.7/4.3
AMP	89	55×10^{-6}	9.7/4.3
MOPA	89	102×10^{-6}	10.1/3.9
CHA	99	440×10^{-6}	10.6/3.4
Carbonic acid	62		6.4/7.6

NOTE: Values given are for atmospheric pressure condensation at 100°C. Higher K_b equates to a lower amine dosage. Carbonic acid pK_d/pK_b values also provided for comparison purposes only.

Neutralizing Amine Summary Notes

- Under particularly difficult conditions, neutralizing amines can be expensive and not always totally successful. The use of satellite dosing points and programs that combine neutralizers with filmers (either fed separately or as a combination product) can produce the desired result. Proof of effectiveness requires pH and Fe/Cu/Ni mass balance profile trials to be undertaken.
- Demand for amine (neutralization capacity) is determined by CO_2 loading and amine equivalent weight. Polyfunctional amines would appear to have a large equivalent weight advantage, but in practice, they often have very low volatilities and poor thermal stability.
- Morpholine or AMP, having low DRs tend to protect the front end of steam/condensate systems, while CHA (with a high DR) protects the furthest regions. DEHA or DMIPA protect the middle regions.
- Rather than calculating CO_2 loading from first principles, a simple measure is to assume a total breakdown evolution of 0.79 ppm CO_2 from every 1 ppm bicarbonate alkalinity, or only 0.35 ppm/1 ppm, if the starting point is carbonate.
- Again, at its simplest allow 5 to 7 ppm of 40% strength amine (any single amine or blended material) for every 1 ppm of CO_2 loading. For most applications, theory seldom matches practice, and the first feed rate calculation should be considered as simply a reference point to establish a base level condensate pH.
- Despite a perceived loss of amine due to carbonic acid neutralization, in practice, much of the amine added is recycled and returned to the FW system. Morpholine tends to suffer a greater recycling loss following passage through the deaerator than either CHA or DEHA. For all amines, loss increases with increase in deaerator operating pressure (from approximately 5% loss of recycled capacity at 5 psig, to nearly 50% at 30 psig. Where amine is applied to FW, the losses are higher than those

facilities where amine is applied overhead (due to BD losses). As operating pressure drops and BD increases, so the trend is for amine losses to be higher.

Some DR values for CO_2, NH_3 and neutralizing amines at various pressures

psig	**0**	**100**	**200**	**300**	**400**	**500**
CO_2	3.0					
NH_3	10.1	7.1	7.1	6.3	5.6	5.3
AMP	0.1	0.47	0.95	0.84	0.85	0.85
CHA	4.0	9.3	23.3	19.5	8.0	6.7
DEA	0.004		0.11			
DEAE	1.7	3.4	4.5	4.4	3.9	3.8
DEHA	1.3					
DMAE	1.0					
DMAEP	1.4					
DMIPA	1.7					
MEA	0.07		0.15			
MOPA	1.0					
Morph	0.4	0.98	1.6	1.4	1.2	1.2
psig	**600**	**700**	**900**	**1000**	**1250**	**1500**
CO_2			99			
NH_3	4.2	4.2	3.9	3.57		
AMP	0.85	0.82	0.91	0.90	0.93	0.92
CHA	6.6	6.1	5.3	4.7	4.4	4.1
DEA				0.07		
DEAE	3.9	3.8	4.5	2.9	2.8	2.6

DEHA						
DMAE	1.9			3.4		
DMAEP						
DMIPA	3.5			3.3		
MEA				0.29		
MOPA	1.9			2.5		
Morph	1.31	1.26	1.24	1.2	1.2	1.2

NOTE: The above data represents results from test work. Some small variation in results is to be expected. Comprehensive data is only available for the most commonly used amines.

Calculating Alkalinity Feed-Rate Requirements

The alkalinity feed-rate requirement is based on FW consumption and calculated as follows:

$$\left[0.67 \times \text{FWCaH} + \left(\text{FWMgH} - \text{FWMA} + \frac{\text{RBWPA}}{\text{COC}}\right)\right] \times \frac{100}{\%\ \text{NaOH in product}}$$

= lb. of product required, per 1×10^6 lb. of boiler FW

Where:

FWCaH = FW calcium hardness, as mg/l (ppm) $CaCO_3$
FWMgH = FW magnesium hardness, as mg/l $CaCO_3$
FWMA = FW methyl orange/total alkalinity, as mg/l $CaCO_3$
RBWPA = Required BW phenolphthalein alkalinity, mg/l $CaCO_3$
COC = Cycles of FW concentration present/desired in the BW

NOTE: If the value within [] brackets is negative, then adequate free alkalinity exists in the FW and the addition of supplementary alkalinity is not required.

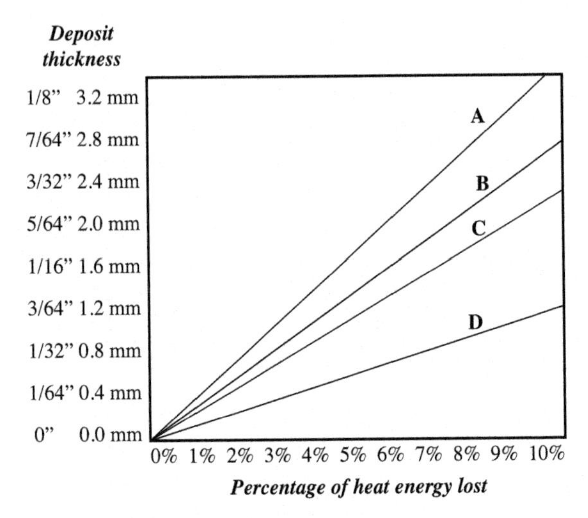

Key: **A** = Typical $CaCO_3$ based boiler sludge
B = Typical deposit including 4 to 5% iron oxides
C = Typical deposit including 8 to 10% iron oxides
D = Typical deposit including 8 to 10% iron and silica

Fig. 10.4 Heat energy losses in WT boiler due to deposition.

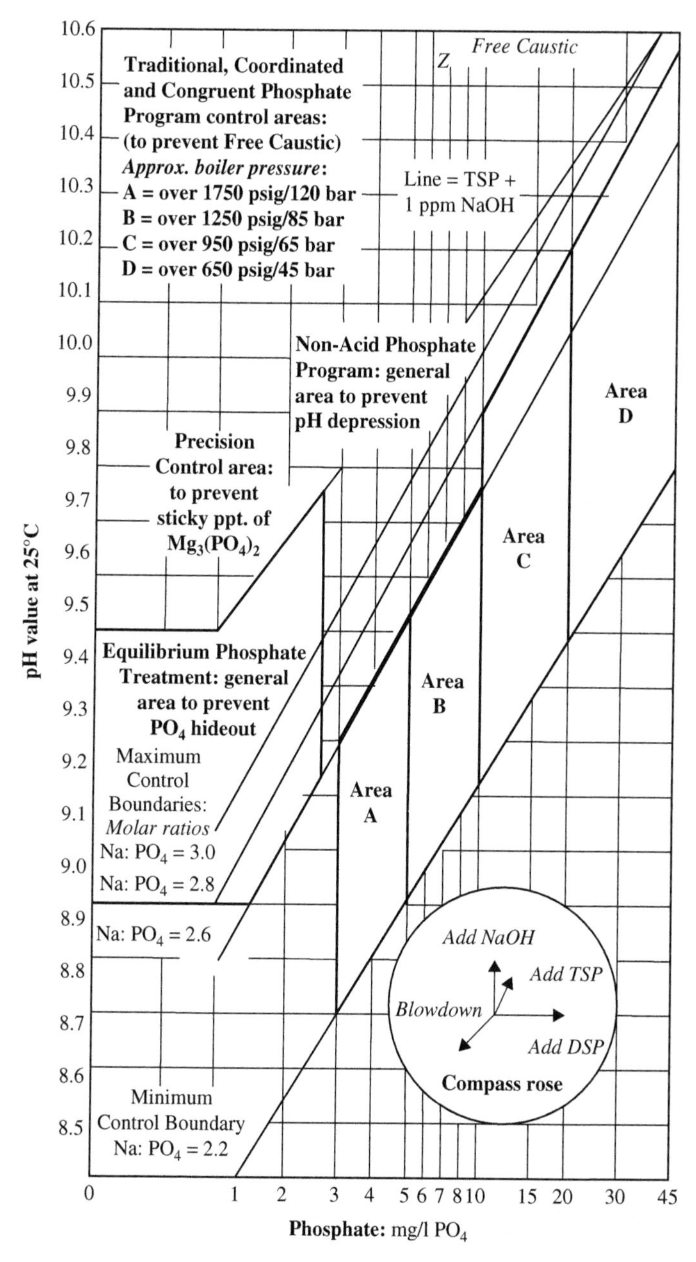

Fig. 10.5 Coordinated, congruent, precision control, equilibrium and non-acid phosphate programs for higher-pressure WT boilers.

ASME Consensus table 1: Suggested water chemistry limits. Industrial watertube, high duty, primary fuel fired, drum typeMakeup water percentage: Up to 100% of feedwater. Conditions: Includes superheater, turbine drives or process restriction on steam purity

Saturated steam purity target: See tabulated values below

Drum Operating Pressure (1) (11) psig (MPa)	**0–300** (0–2.07)	**301–450** (2.08–3.10)	**451–600** (3.11–4.14)	**601–750** (4.15–5.17)	**751–900** (5.18–6.21)	**901–1000** (6.22–6.89)	**1001–1500** (6.90–10.34)	**1501–2000** (10.35–13.79)
Feedwater (7) Dissolved oxygen ppm (mg/l) O_2 – measured before chemical oxygen scavenger addition (8)	<0.007	<0.007	<0.007	<0.007	<0.007	<0.007	<0.007	<0.007
Total iron ppm (mg/l) Fe	≤0.1	≤0.05	≤0.03	≤0.025	≤0.02	≤0.02	≤0.01	≤0.01
Total copper ppm (mg/l) Cu	≤0.05	≤0.025	≤0.02	≤0.02	≤0.015	≤0.01	≤0.01	≤0.01
Total hardness ppm as $CaCO_3$	≤0.3	≤0.3	≤0.2	≤0.2	≤0.1	≤0.05	ND	ND
pH @ 25°C	8.3–10.0	8.3–10.0	8.3–10.0	8.3–10.0	8.3–10.0	8.8–9.6	8.8–9.6	8.8–9.6
Chemicals for preboiler system protection	NS	NS	NS	NS	NS	VAM	VAM	VAM
Nonvolatile TOC ppm C (6)	<1	<1	<0.5	<0.5	<0.5	<0.2	<0.2	<0.2
Oily matter ppm (mg/l)	<1	<1	<0.5	<0.5	<0.5	<0.2	<0.2	<0.2

Boiler Water								
Silica ppm (mg/l) SiO_2	≤150	≤90	≤40	≤30	≤20	≤8	≤2	≤1
Total alkalinity ppm as $CaCO_3$	<350 (3)	<300 (3)	<250 (3)	<200 (3)	<150 (3)	<100 (3)	NS (4)	NS (4)
Free OH alk. ppm $CaCO_3$ (2)	NS	NS	NS	NS	NS	NS	ND (4)	ND (4)
Specific conductance (12) μmhos/cm (μS/cm) @ 25°C without neutralization	5400–1100 (5)	4600–900 (5)	3800–800 (5)	1500–300 (5)	1200–200 (5)	1000–200 (5)	≤150	≤80
TDS in Steam (9)								
TDS (maximum) ppm (mg/l)	1.0–0.2	1.0–0.2	1.0–0.2	0.5–0.1	0.5–0.1	0.5–0.1	0.1	0.1

NOTES: NS = not specified, ND = not detectable, VAM = Use only volatile alkaline materials upstream of attemperation water source (10).

ASME Consensus table 2: Suggested chemistry limits. Industrial watertube, high duty, primary fuel fired, drum type

Makeup water percentage: Up to 100% of feedwater.
Conditions: No superheater, turbine drives, or process restriction on steam purity
Steam purity (7)**:** 1.0 ppm (mg/l) TDS maximum

Drum Operating Pressure psig (MPa)	**0–300** (0–2.07)	**301–600** (2.08–4.14)
Feedwater (3) Dissolved oxygen ppm (mg/l) O_2 – measured before chemical oxygen scavenger addition (1) (2)	<0.007	<0.007
Total iron ppm (mg/l) Fe	<0.1	<0.05
Total copper ppm (mg/l) Cu	<0.05	<0.025
Total hardness ppm (mg/l) as $CaCO_3$	<0.5	<0.3
pH @ 25°C	8.3 to 10.5	8.3 to 10.5
Nonvolatile TOC ppm (mg/l) C (6)	<0.1	<0.1
Oily matter ppm (mg/l)	<0.1	<0.1
Boiler Water Silica ppm (mg/l) SiO_2	<150	<90
Total alkalinity ppm as $CaCO_3$	<1000 (5)	<850 (5)
Free OH alkalinity ppm as $CaCO_3$ (4)	not specified	not specified
Specific conductance µmhos/cm (µS/cm) @ 25°C without neutralization	<7000 (5)	<5500 (5)

ASME Consensus table 3: Suggested chemistry limits. Industrial firetube, high duty, primary fuel fired

Makeup water percentage: Up to 100% of feedwater.
Conditions: No superheater, turbine drives, or process restriction on steam purity
Steam purity (7)**:** 1.0 ppm (mg/l) TDS maximum

Drum Operating Pressure psig (MPa)	**0–300** (0–2.07)
Feedwater (3) Dissolved oxygen ppm (mg/l) O_2 – measured before chemical oxygen scavenger addition (1) (2)	<0.007
Total iron ppm (mg/l) Fe	<0.1
Total copper ppm (mg/l) Cu	<0.05
Total hardness ppm as $CaCO_3$	<1.0
pH @ 25°C	8.3 to 10.5
Nonvolatile TOC ppm (mg/l) C (6)	<10
Oily matter ppm (mg/l)	<0.1
Boiler Water Silica ppm (mg/l) SiO_2	<150
Total alkalinity ppm as $CaCO_3$	<700 (5)
Free OH alkalinity ppm as $CaCO_3$ (4)	not specified
Specific conductance µmhos/cm (µS/cm) @ 25°C without neutralization	<7000 (5)

ASME Consensus table 4: Suggested water chemistry limits. Industrial coil type, watertube, high duty, primary fuel fired rapid steam generators

Makeup water percentage: Up to 100% of water to the coil.
Steam to water ratio (volume to volume): Up to 4000:1
Total evaporation: Up to 95% of the water to the coil.
Saturated steam purity target: See tabulated values below

Drum Operating Pressure psig (MPa)	**0–300** (0–2.07)	**301–450** (2.08–3.10)	**451–600** (3.11–4.14)	**601–900** (4.15–6.21)	**>900** (>6.21)
Steam Purity Targets (1) Specific conductance µmhos/cm (µS/cm) @ 25°C	≤50 (2)	≤24 (2)	≤20 (2)	≤0.5	≤0.2
Dissolved solids ppm (mg/l)	≤25	≤12	≤10	≤0.25	≤0.01
Silica ppm (mg/l) SiO_2	NS	NS	NS	≤0.03	≤0.02
Water to coil (3) Dissolved oxygen ppm (mg/l) O_2 – measured after chemical oxygen scavenger addition (4)	<0.007	<0.007	<0.007	<0.007	<0.007

Total iron ppm (mg/l) Fe	<1.0	<0.3	<0.1	≤0.05	≤0.02
Total copper ppm (mg/l) Cu	<0.1	<0.05	<0.03	≤0.02	≤0.02
Total hardness ppm (mg/l) as $CaCO_3$	0–trace	0–trace	0–trace	ND (6)	ND (6)
pH @ 25°C	9.0–11.0	9.0–11.0	9.0–11.0	9.0–11.0	9.0–11.0
Total alkalinity ppm (mg/l) as $CaCO_3$	<800	<600	<500	<200	<100 (7)
Hydroxide alkalinity ppm (mg/l) as $CaCO_3$ (5)	<300	<200	<120	<60	≤50 (7)
Silica ppm (mg/l) SiO_2	≤150	≤100	≤60	≤30	≤10 (7)
Specific conductance µmhos/cm (µS/cm) @ 25°C without neutralization	<8000	<6000	<5000	<4000	<500 (7)

NOTE: NS = not specified, ND = not detectable

ASME Consensus table 5: Suggested water chemistry limits. Marine propulsion, watertube, oil fired drum type

Makeup water percentage: Up to 5% of feedwater
Pretreatment: At sea, evaporator condensate; in port, evaporator condensate or water from shore facilities meeting feedwater quality guidelines.
Saturated steam purity (6)**:** 30 ppb (µg/l) TDS max., 10 ppb (µg/l) Na max., 20 ppb (µg/l) SiO_2 max.

Drum Operating Pressure psig (MPa)	**450–850** (3.1–5.86)	**851–1250** (5.87–8.62)
Feedwater (1) Dissolved oxygen ppm (mg/l) O_2 – measured before chemical oxygen scavenger addition (5)	<0.007	<0.007
Total iron ppm (mg/l) Fe	<0.02	<0.01
Total copper ppm (mg/l) Cu	<0.01	<0.005
Total hardness ppm (mg/l) as $CaCO_3$	<0.1	<0.05
pH @ 25°C	8.3–9.0	8.3–9.0
Chemical for preboiler system protection	VAM	VAM
Oily matter ppm (mg/l)	<0.05	<0.05
Boiler Water Silica ppm (mg/l) SiO_2	<30	<5
Total alkalinity ppm (mg/l) as $CaCO_3$ (4)	NS (4)	NS (4)
OH alkalinity ppm (mg/l) as $CaCO_3$ (4)	<200 (3)	ND (4)
Specific conductance µmhos/cm (µS/cm) @ 25°C without neutralization (2)	<700	<150

NOTE: NS = not specified, ND = not detectable, VAM = use only volatile alkaline materials

ASME Consensus table 6: Suggested water chemistry limits. Electrode, high voltage, forced circulation jet type

Makeup water percentage: Up to 100% of feedwater
Conditions: No superheater, turbine drives, or process restriction on steam purity

Operating Pressure psig (MPa)	**0–450** (0–3.1)
Feedwater (2) Dissolved oxygen ppm (mg/l) O_2 – measured before chemical oxygen scavenger addition (1)	<0.007
Total hardness ppm (mg/l) as $CaCO_3$	<0.25
pH @ 25°C	8.3–10.5
Nonvolatile TOC ppm (mg/l) C (6)	NS (8)
Boiler Water pH @ 25°C	8.3–10.5
Silica ppm (mg/l) SiO_2	<150
Total alkalinity ppm (mg/l) as $CaCO_3$	<350 (3)
OH alkalinity ppm (mg/l) as $CaCO_3$ (8)	NS (8)
Total iron ppm (mg/l) Fe plus total copper ppm (mg/l) Cu	2.0 (2) (7)
Suspended solids	NS (7)
Organic matter	NS (8)
Specific conductance μmhos/cm (μS/cm) @ 25°C without neutralization	NS (5)

NOTE:

1. Tables extracted from ASME Consensus are reproduced above with kind permission of The American Society of Mechanical Engineers, New York, NY, USA

Reduction in heat transfer efficiency with increase in deposit thickness, where **A** = iron oxides and silica, **B** = iron oxides, **C** = calcium carbonate.

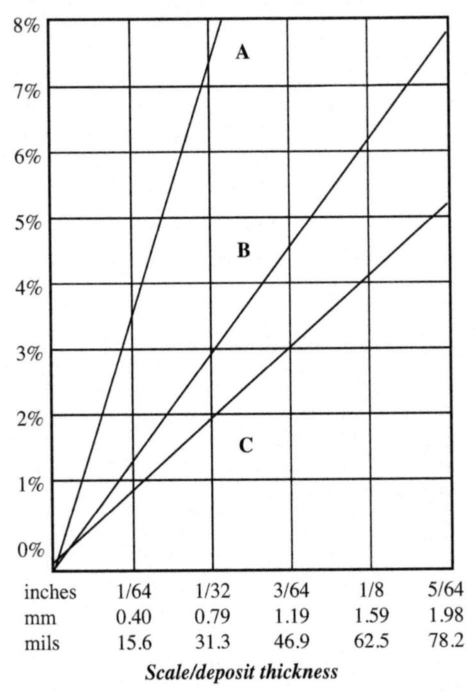

Notes

CPSIA information can be obtained
at www.ICGtesting.com
Printed in the USA
BVOW11s1522200316
441001BV00003B/11/P